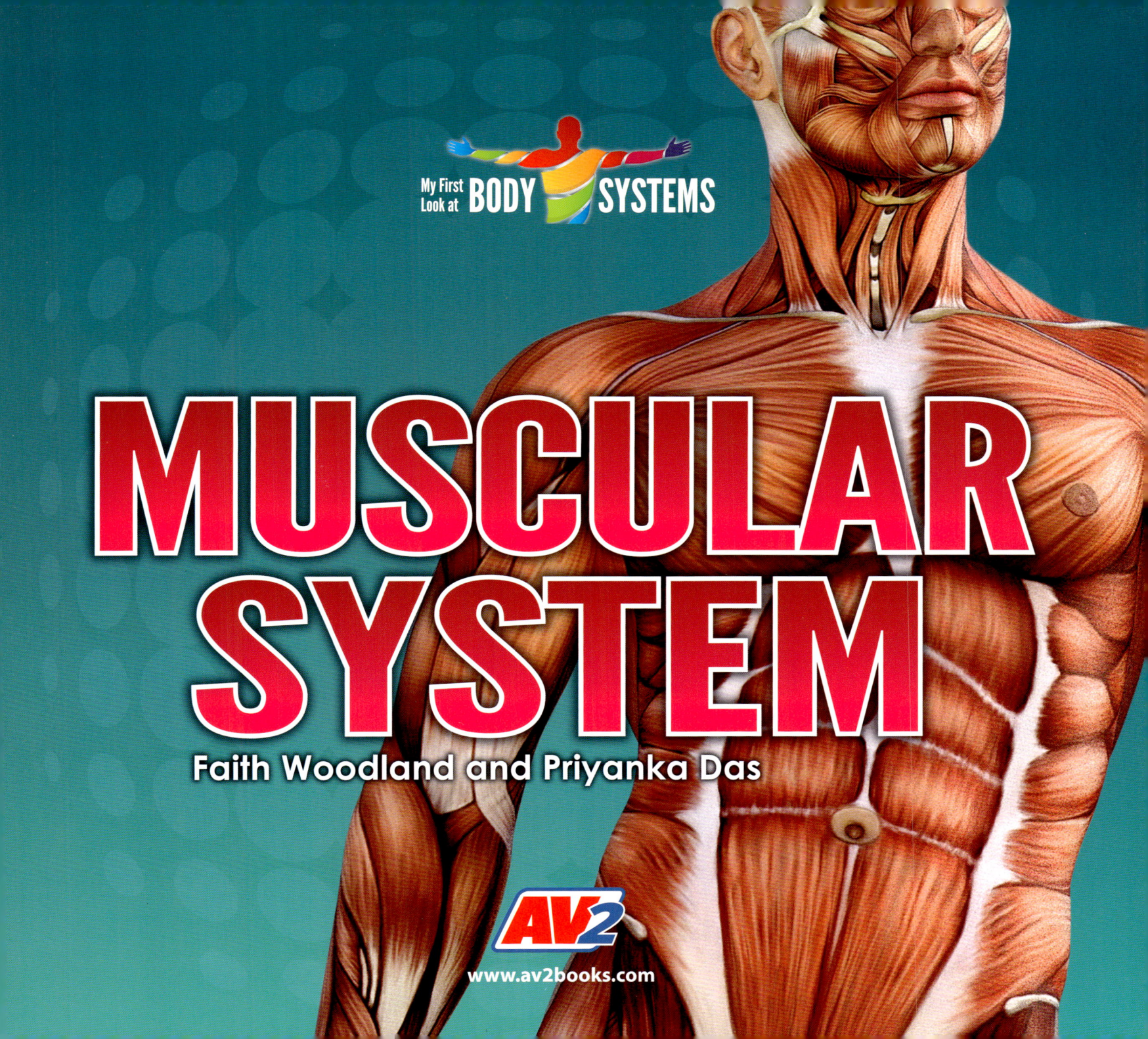

My First Look at BODY SYSTEMS

MUSCULAR SYSTEM

Faith Woodland and Priyanka Das

AV2
www.av2books.com

Step 1
Go to **www.av2books.com**

Step 2
Enter this unique code

ZKUDX2M9Y

Step 3
Explore your interactive eBook!

Your interactive eBook comes with...

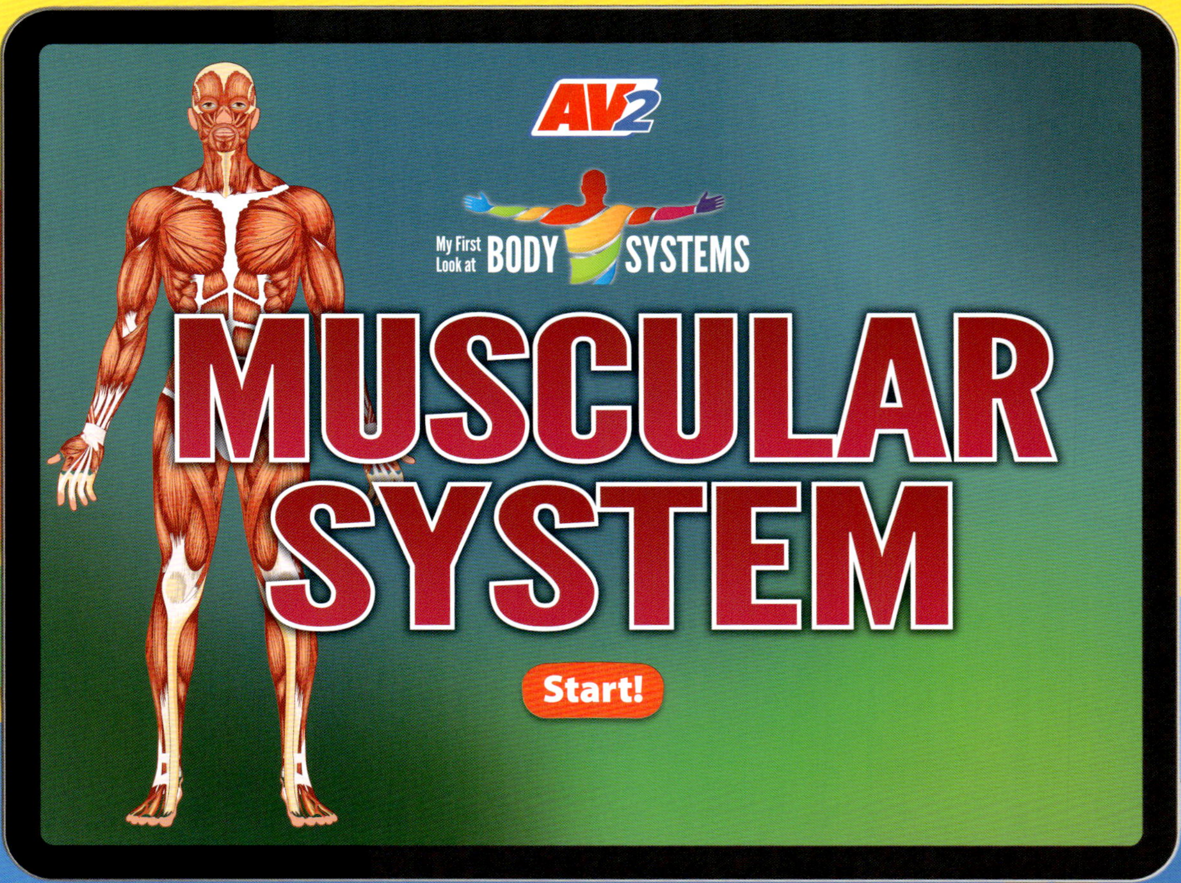

AV2 is optimized for use on any device

Audio
Listen to the entire book read aloud

Videos
Watch informative video clips

Weblinks
Gain additional information for research

Try This!
Complete activities and hands-on experiments

Key Words
Study vocabulary, and complete a matching word activity

Quizzes
Test your knowledge

Slideshows
View images and captions

View new titles and product videos at www.av2books.com

2

MUSCULAR SYSTEM

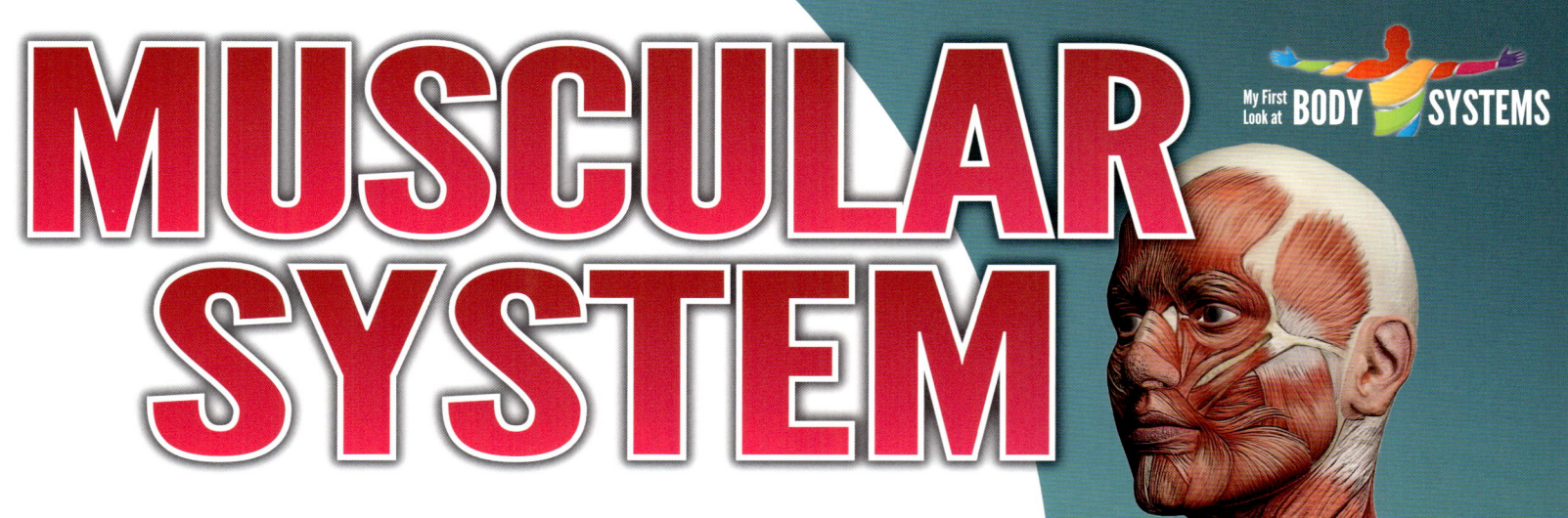

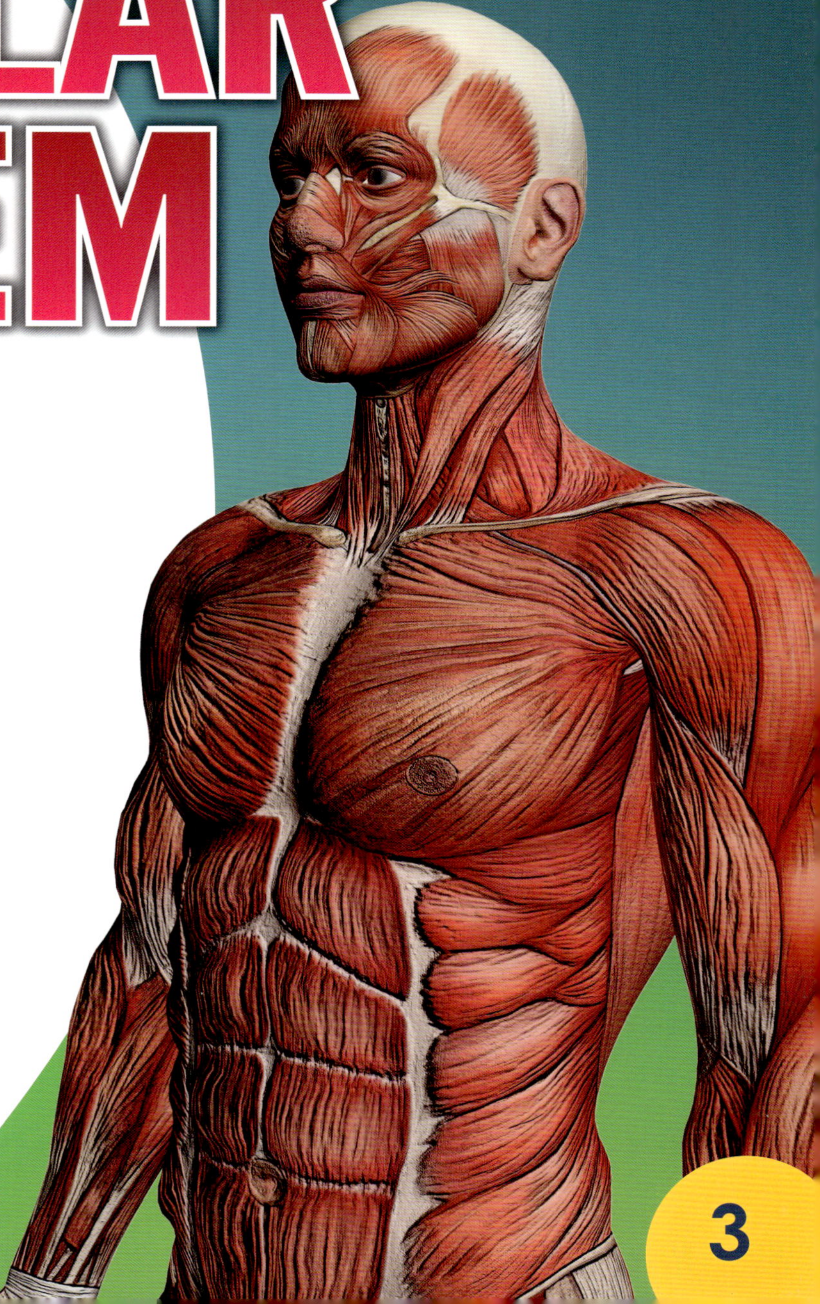

CONTENTS

- 2   AV2 Book Code
- 4   Muscular System
- 6   All about Muscles
- 8   Parts of the Muscular System
- 10   Cardiac Muscles
- 12   Smooth Muscles
- 14   Skeletal Muscles
- 16   Tendons
- 18   Staying Healthy
- 20   Career Spotlight
- 22   Muscular System Quiz
- 24   Key Words

Muscular System

The human body is made up of many systems. These systems work together to keep us healthy and alive. **The muscular system is the body system that makes movement possible.**

The muscular system is always working. **You use it every time you get up and move around.** The muscular system even works when you are resting.

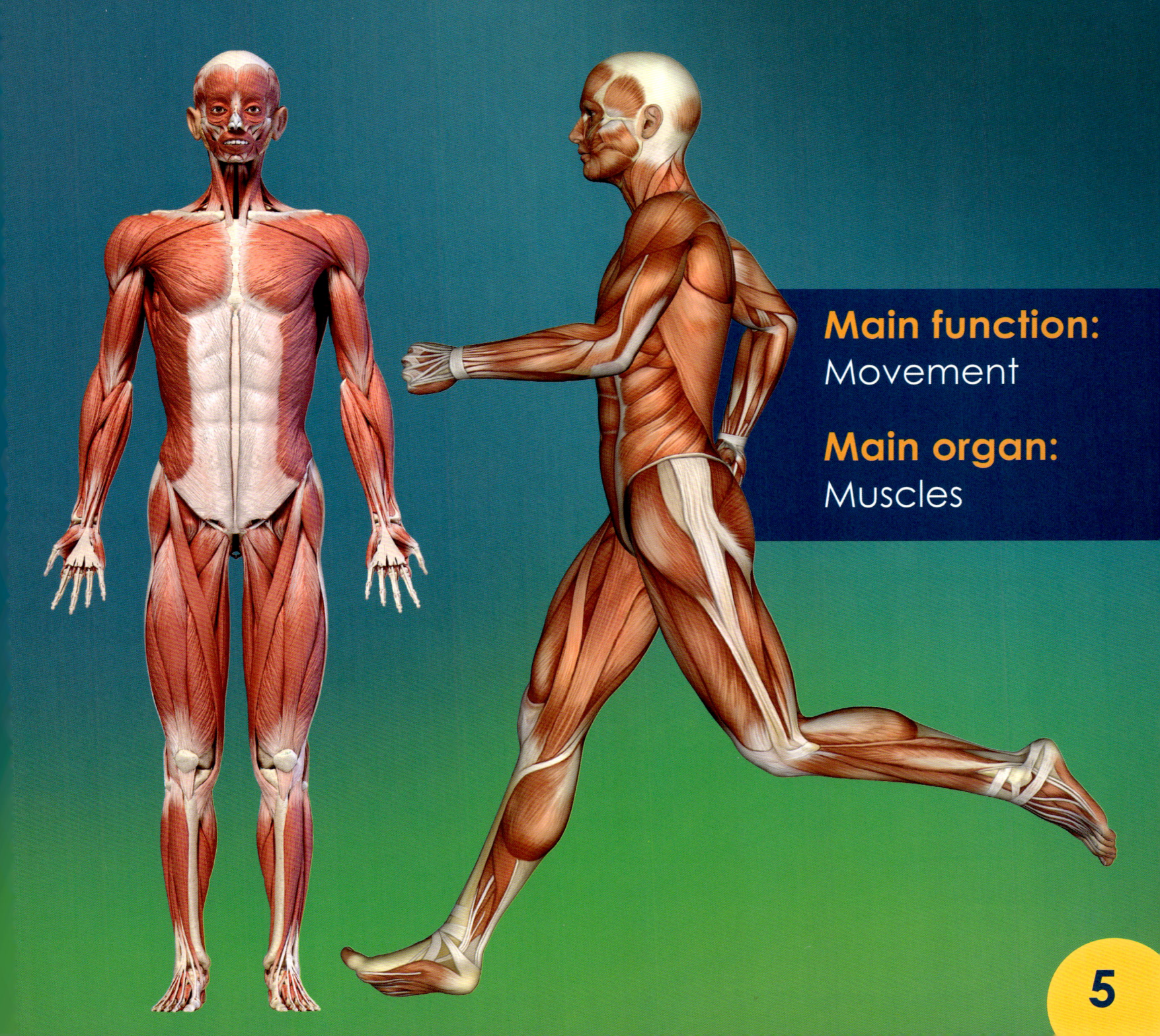

All about Muscles

Muscles are the main organs of the muscular system. **They are a little like rubber bands.** They tighten and relax. This causes movement.

Some muscles work without you even thinking about them. Others work when you want them to. Muscles let you run, jump, and swim. They allow you to lift objects. Every movement of your body is possible because of muscles.

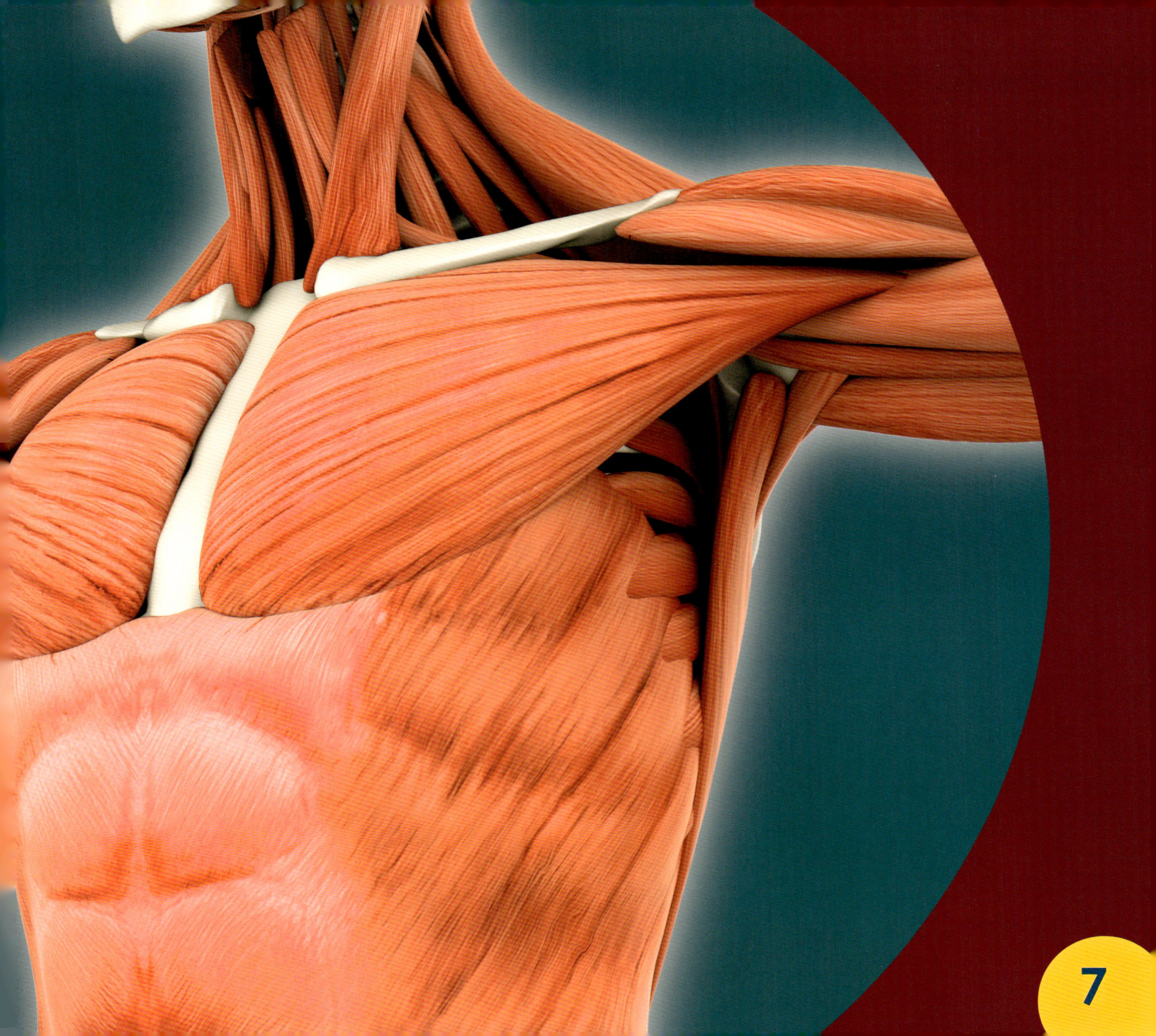

Parts of the Muscular System

There are three main types of muscles. They are cardiac, smooth, and skeletal muscles.

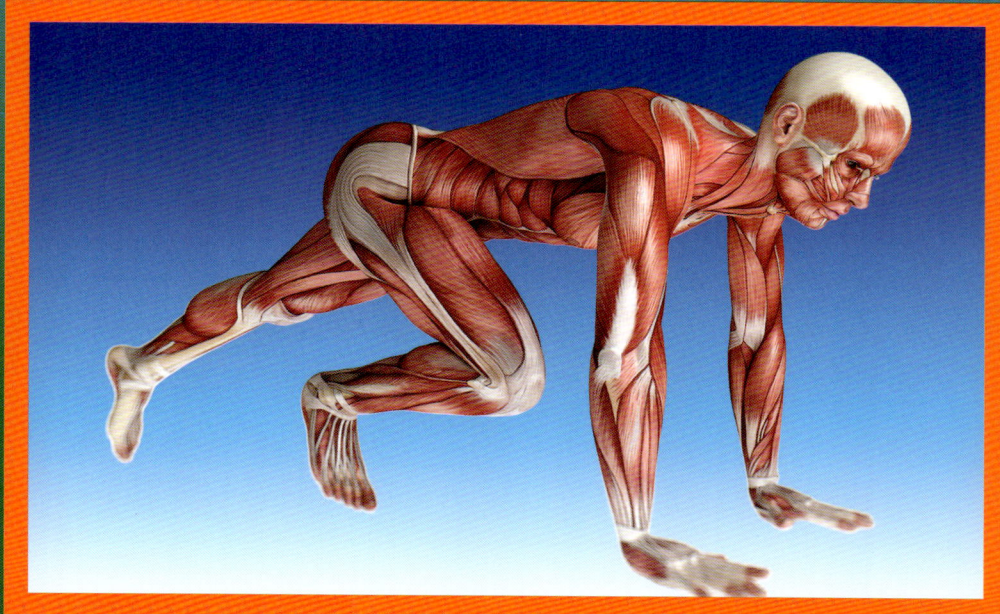

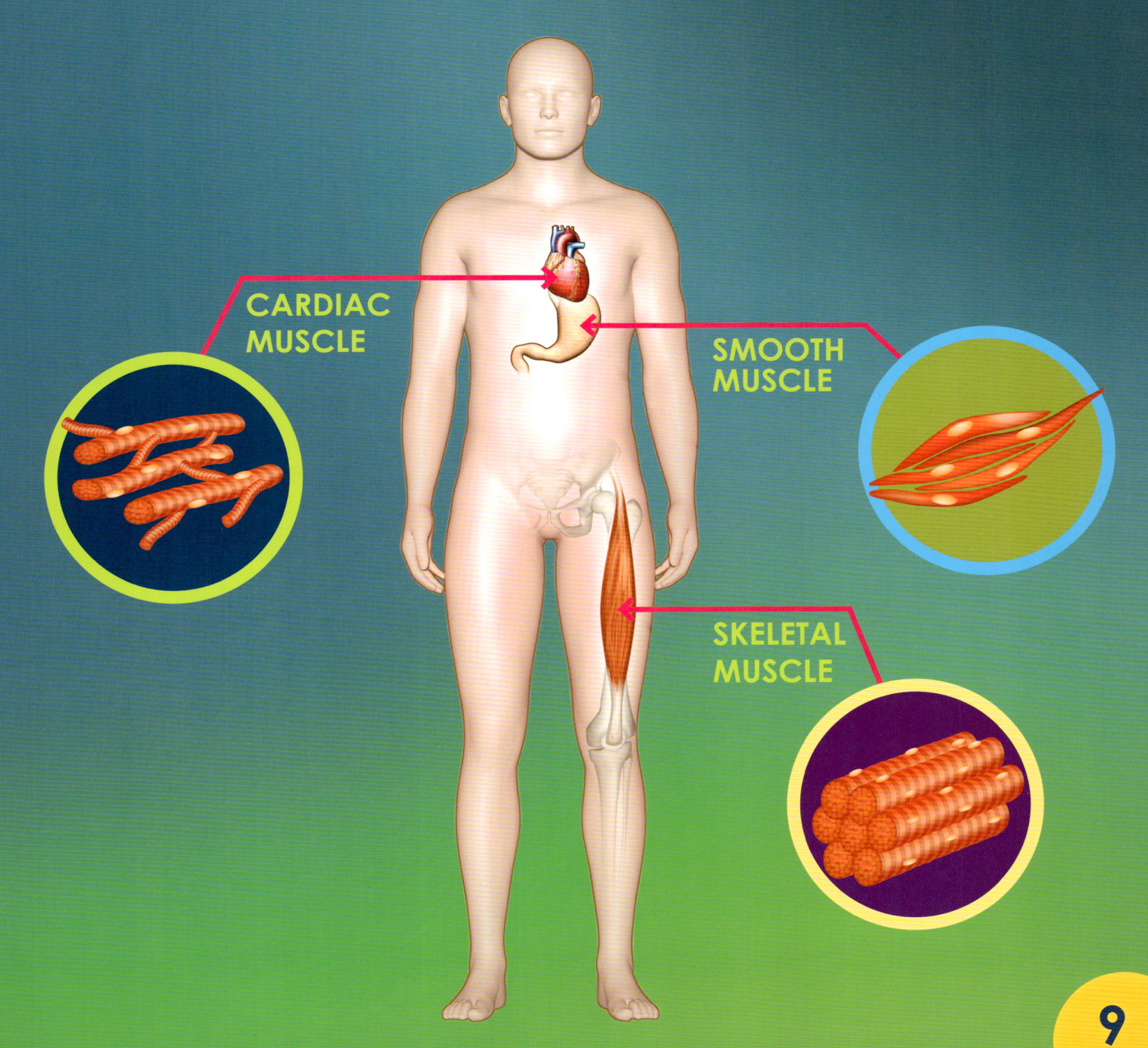

Cardiac Muscles

Cardiac muscles are a special type of muscle. **They are only found in the heart.** These muscles keep the heart beating.

When cardiac muscles tighten, they make the heart pump blood. This blood travels through the body's circulatory system.

The heart is the **hardest working muscle** in the body. It can beat more than **3 billion times** in a person's life.

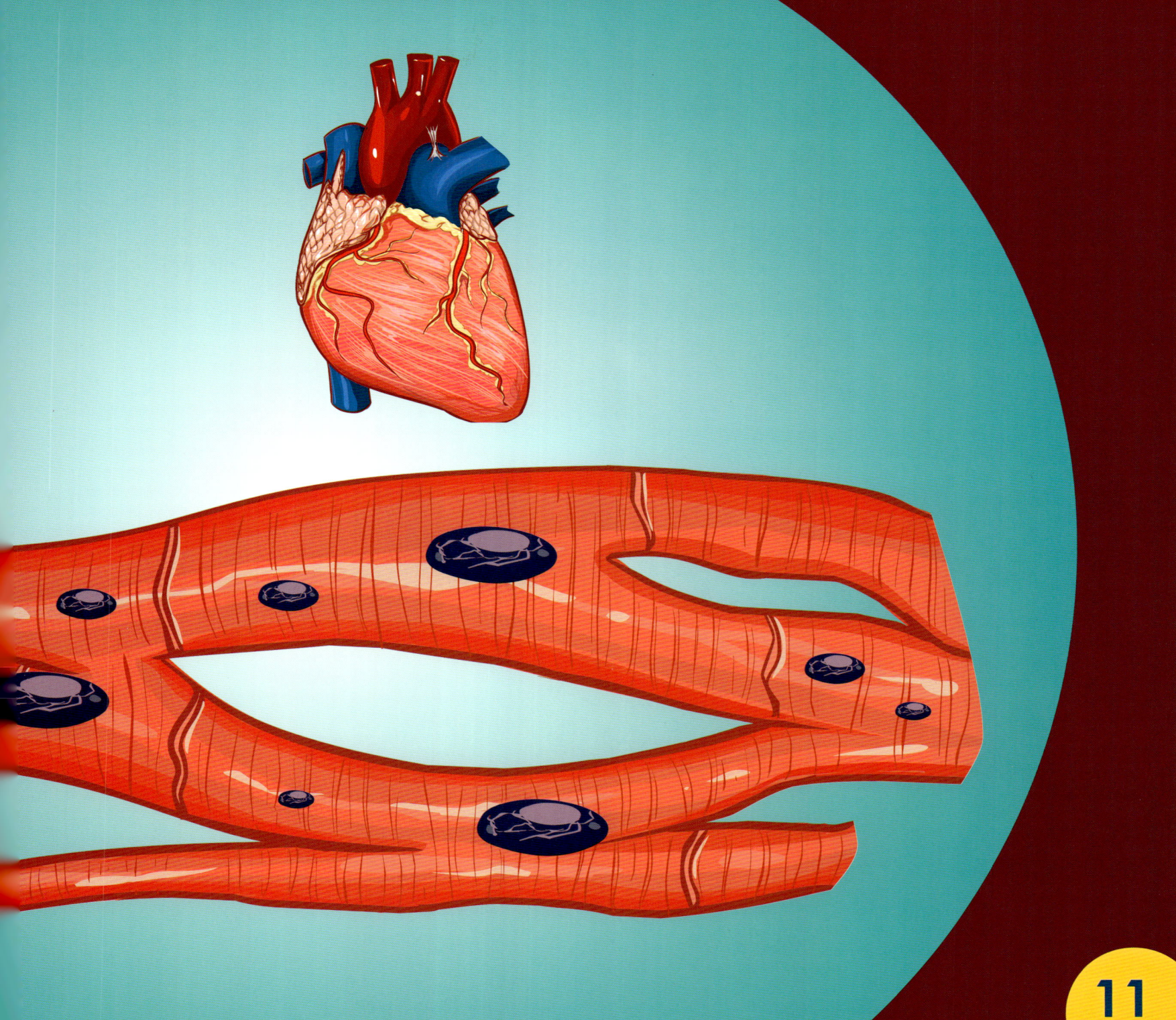

Smooth Muscles

Like cardiac muscles, smooth muscles work without you thinking about them. They are found in the walls of some organs. **Smooth muscles help body systems with movement inside the body.**

Smooth muscles can be found in the walls of the stomach and intestines. **The muscles push food through the digestive system when they tighten.** Smooth muscles are in the walls of blood vessels, too. They tighten to help the circulatory system control blood flow.

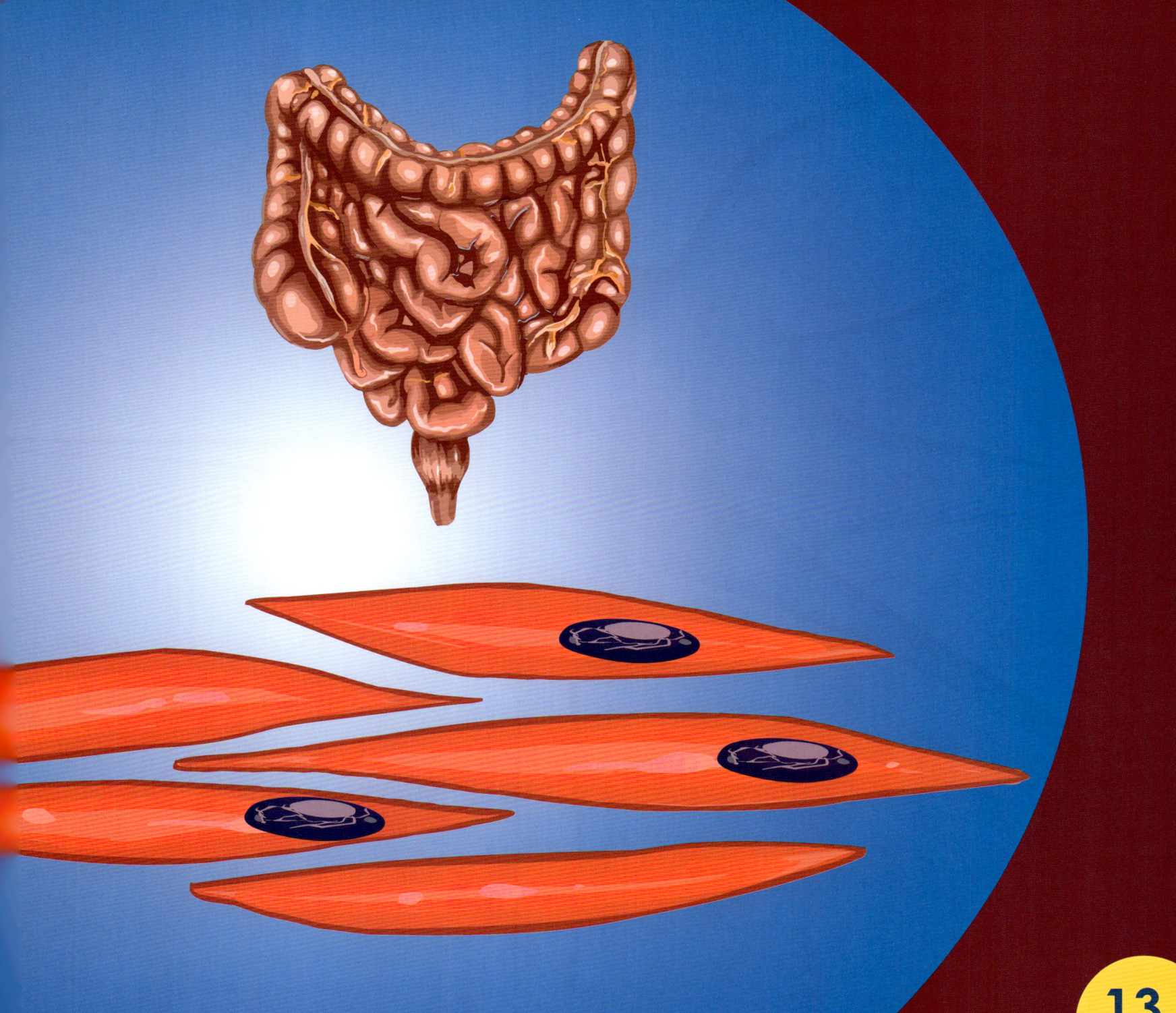

Skeletal Muscles

When most people think of muscles, they picture skeletal muscles. **Skeletal muscles are the only muscles in the body that you can choose to move.**

Skeletal muscles are attached to the skeleton. **They work with the bones of the skeletal system.** Try lifting your hand. Move your leg a little, or wiggle your toes. Every time you move, you are using skeletal muscles.

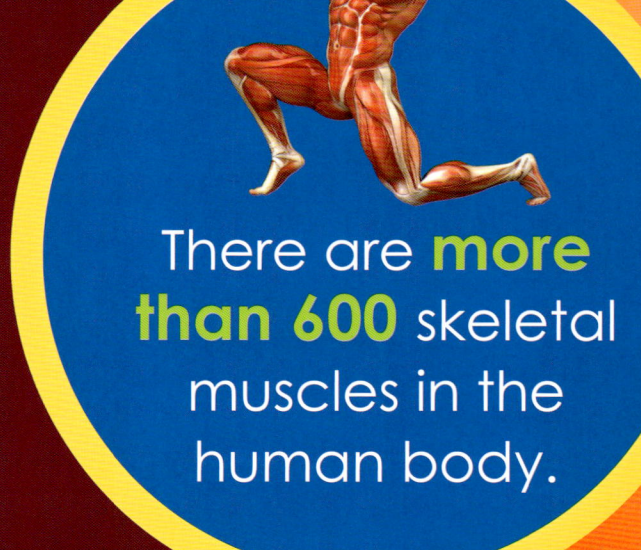

There are **more than 600** skeletal muscles in the human body.

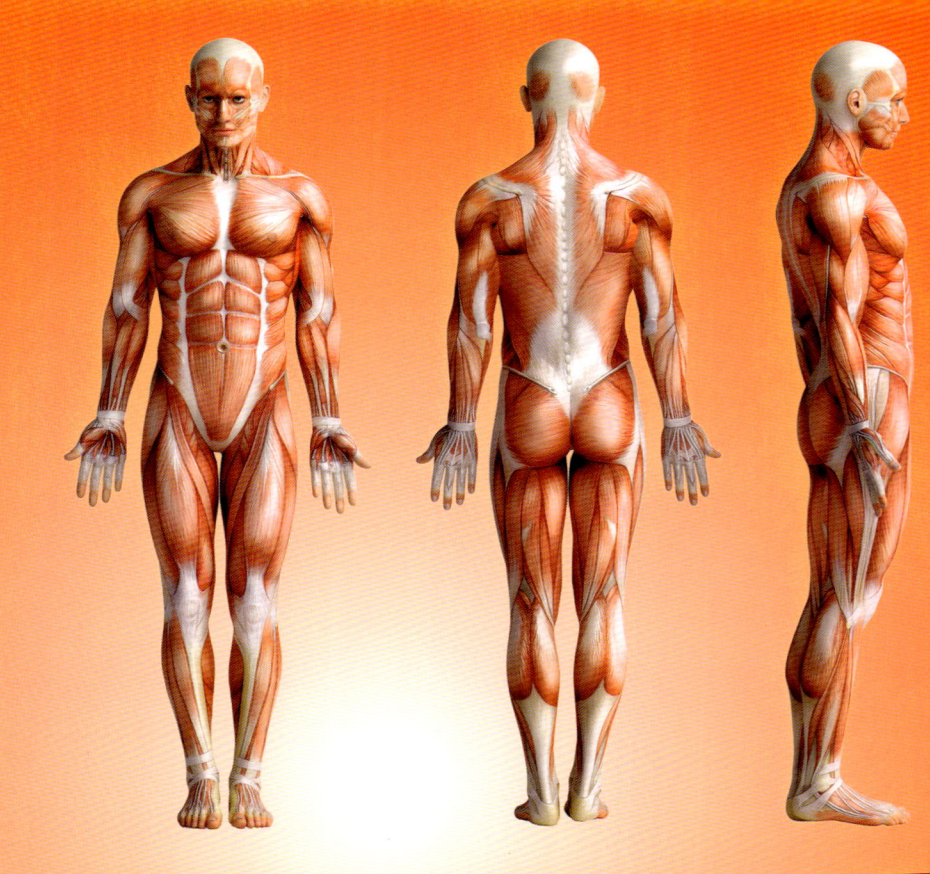
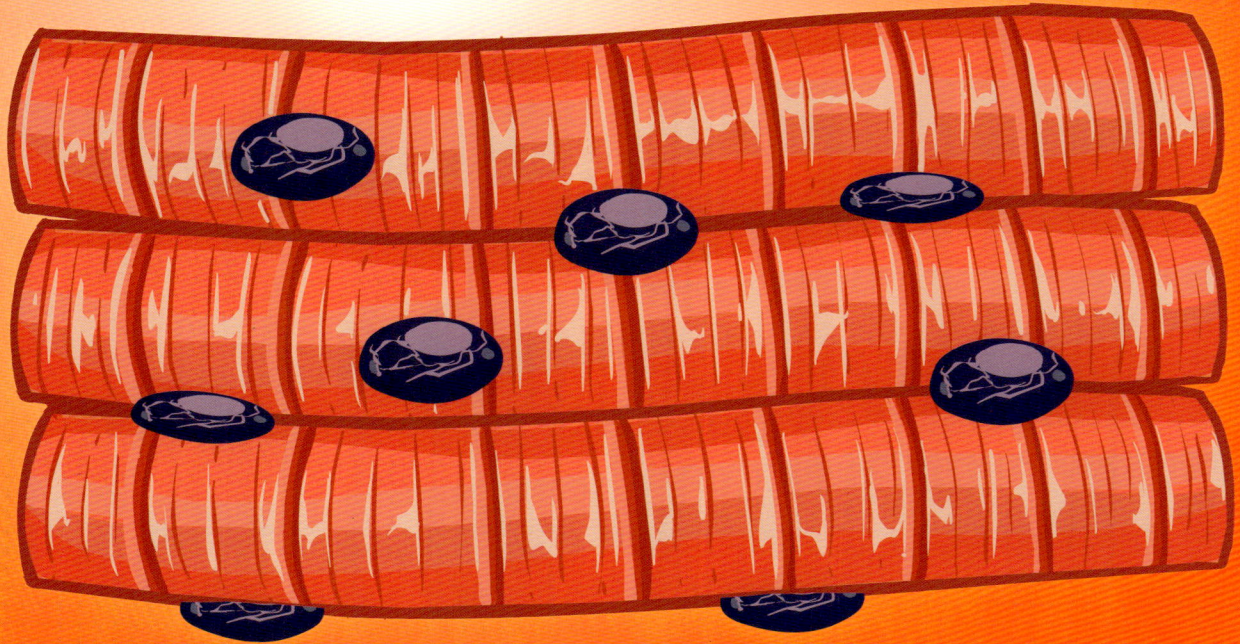

15

Tendons

Tendons connect muscles to bones. When a muscle tightens, the tendon pulls the attached bone. The bone moves. Tendons can also connect muscles to other muscles.

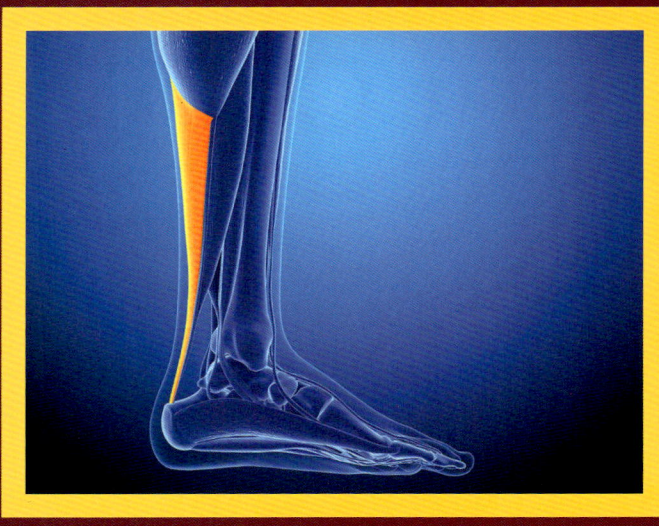

The largest tendon in the body is the Achilles tendon. It is in the leg. This tendon connects muscles in the lower leg to a bone in the foot.

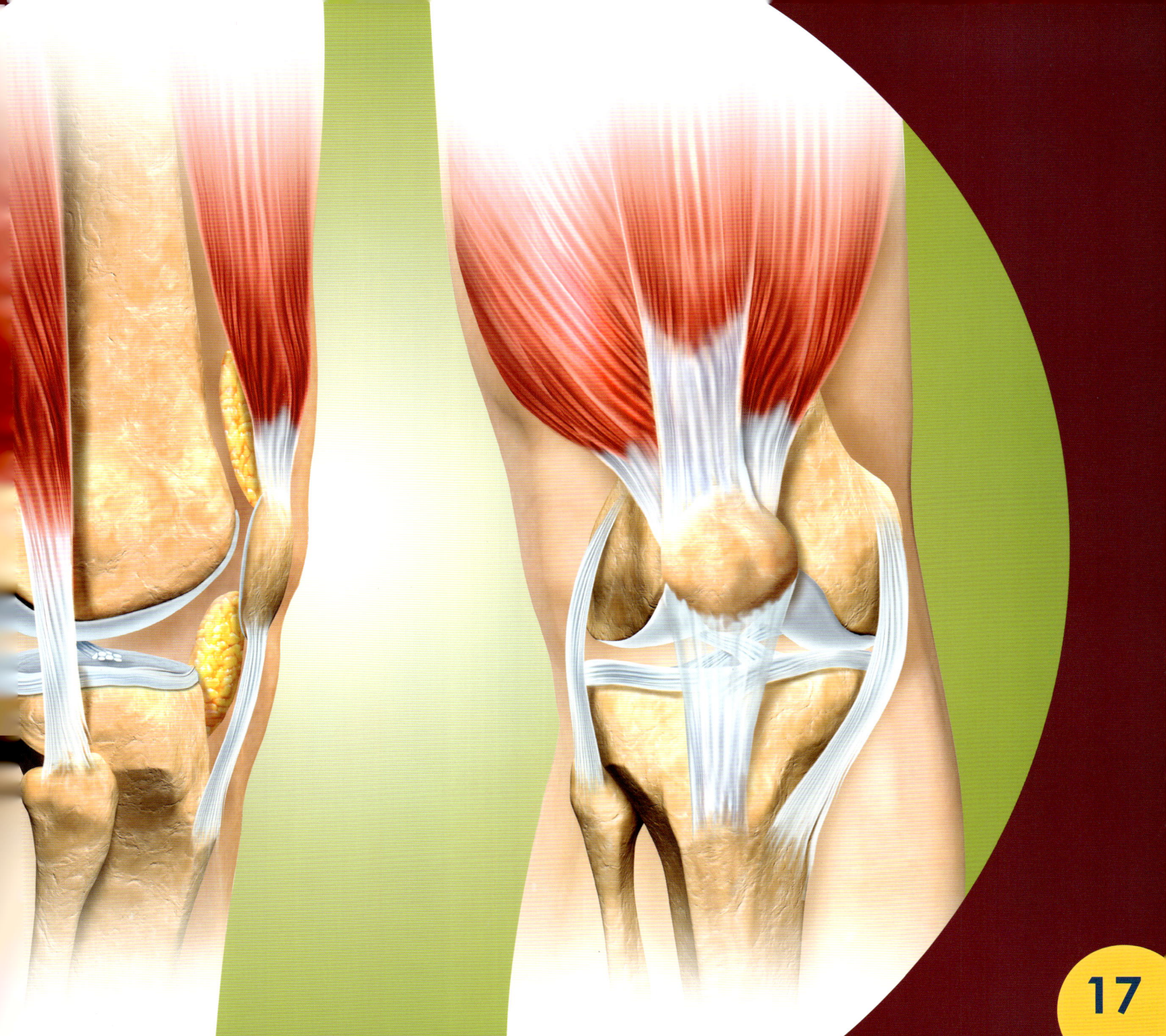

Staying Healthy

Protein is an important nutrient. **Foods that are high in protein help build muscle.** Meat, fish, eggs, and nuts are high in protein.

Exercise is very good for the muscular system. Lifting weights strengthens muscles. It helps them grow. Exercises such as running and jumping improve stamina. Stretching keeps the muscles flexible.

lifting weights

jumping

running

stretching

Career Spotlight

Fitness instructors help people keep the muscular system active. They lead people in exercise activities. Fitness instructors make exercise plans that improve the strength of the muscular system. They may also help people decide what foods to eat to keep the body healthy.

ARIZONA

The National Academy of Sports Medicine in Arizona trains many fitness instructors in the United States.

Muscular System Quiz

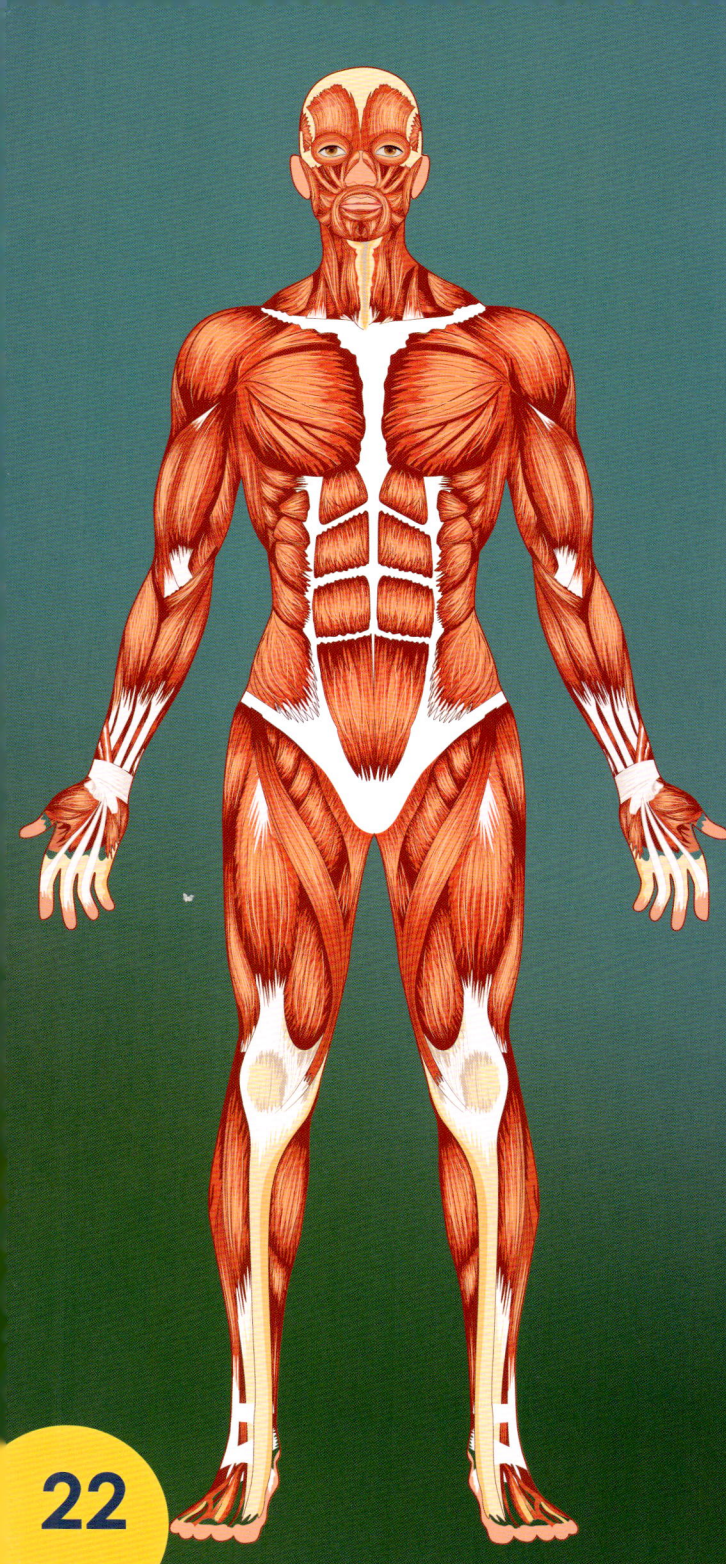

Can you name the parts of the muscular system shown in these pictures?

A] Cardiac muscles

B] Smooth muscles

C] Skeletal muscles

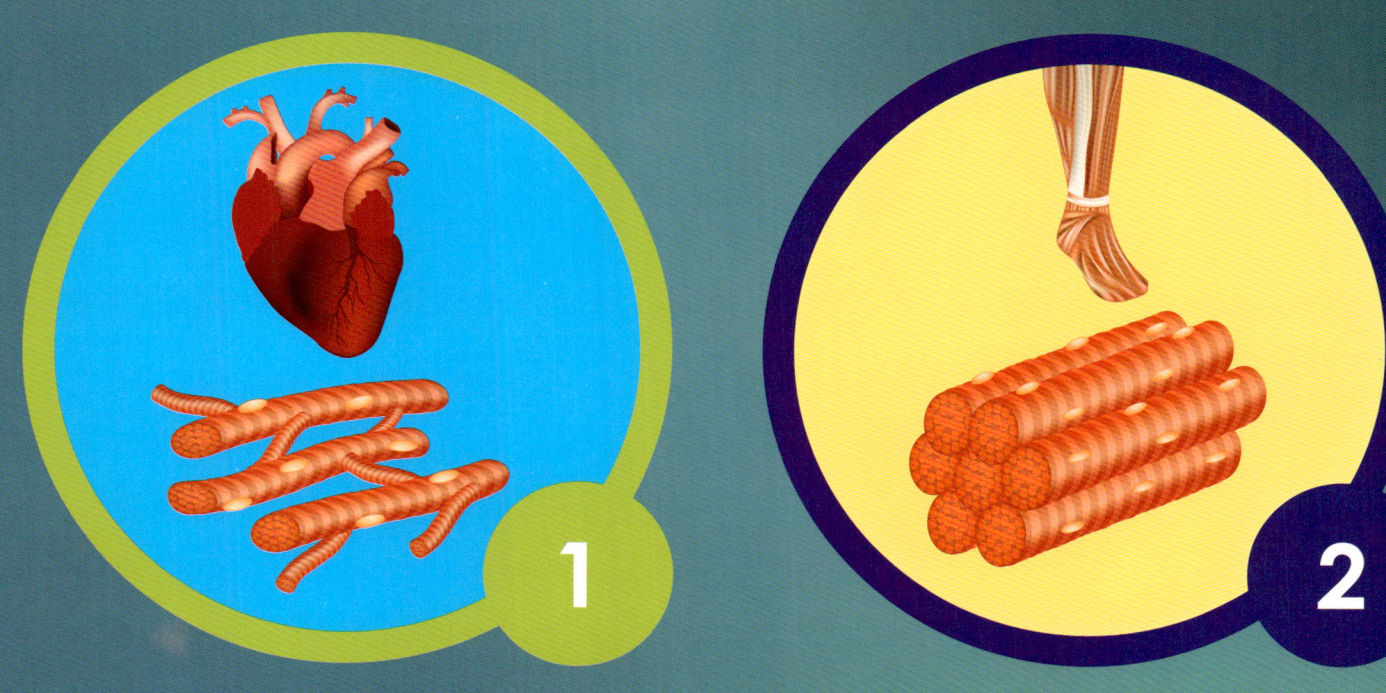

Answers
1] A
2] C
3] B

23

KEY WORDS

Research has shown that as much as 65 percent of all written material published in English is made up of 300 words. These 300 words cannot be taught using pictures or learned by sounding them out. They must be recognized by sight. This book contains 78 common sight words to help young readers improve their reading fluency and comprehension. This book also teaches young readers several important content words, such as proper nouns. These words are paired with pictures to aid in learning and improve understanding.

Page	Sight Words First Appearance
4	always, and, are, around, even, every, get, is, it, keep, made, makes, many, move, of, that, the, these, time, to, together, up, us, use, when, work, you
6	a, about, all, because, let, like, little, others, run, some, them, they, this, want, without, your
8	parts, there, three
10	can, found, in, life, more, only, than, through
12	be, food, help, too, with
14	hand, most, or, people, picture, think, try
16	also
18	an, high, important
19	as, for, good, such, very
20	may, states, what

Page	Content Words First Appearance
4	human body, movement, muscular system
5	function, organ, muscles
6	objects, rubber bands
8	cardiac muscle, skeletal muscle, smooth muscle, types
10	blood, circulatory system, heart, person
12	blood vessels, flow, intestines, stomach, walls
14	bones, leg, toes, skeletal system, skeleton
16	Achilles tendon, foot, tendons
18	eggs, fish, meat, nutrient, nuts, protein
19	exercise, stamina, weights
20	Arizona, activities, fitness instructors, National Academy of Sports Medicine, plans, strength, United States

Published by AV2
14 Penn Plaza, 9th Floor New York, NY 10122
Website: www.av2books.com

Copyright ©2021 AV2
All rights reserved. No part of this publication may be reproduced, stored in a retrieval system, or transmitted in any form or by any means, electronic, mechanical, photocopying, recording, or otherwise, without the prior written permission of the publisher.

Library of Congress Cataloging-in-Publication Data available upon request

ISBN 978-1-7911-1880-8 (hardcover)
ISBN 978-1-7911-1881-5 (softcover)
ISBN 978-1-7911-1882-2 (multi-user eBook)
ISBN 978-1-7911-1883-9 (single-user eBook)

Printed in Guangzhou, China
1 2 3 4 5 6 7 8 9 0 24 23 22 21 20

062020
100919

Project Coordinator: Priyanka Das   Designer: Jean Faye Marie Rodriguez

Every reasonable effort has been made to trace ownership and to obtain permission to reprint copyright material. The publisher would be pleased to have any errors or omissions brought to its attention so that they may be corrected in subsequent printings.

The publisher acknowledges iStock and Shutterstock as the primary image suppliers for this title.